Modelling with circular motion

Unit Guide

The School Mathematics Project

Main authors	Stan Dolan
	Judith Galsworthy
	Janet Jagger
	Ann Kitchen
	Paul Roder
	Mike Savage
	Bernard Taylor
	Carole Tyler
	Nigel Webb
	Phil Wood
Team leader	Ann Kitchen
Project director	Stan Dolan

This unit has been produced in collaboration with the Mechanics in Action Project, based at the Universities of Leeds and Manchester.

The authors would like to give special thanks to Ann White for her help in producing the trial edition and in preparing this book for publication

The publishers would like to thank K.E Kitchen for supplying the photograph on page (ix).

Published by the Press Syndicate of the University of Cambridge
The Pitt Building, Trumpington Street, Cambridge CB2 1RP
40 West 20th Street, New York, NY 10011–4211, USA
10 Stamford Road, Oakleigh, Victoria 3166, Australia

© Cambridge University Press 1992

First published 1992

Produced by Gecko Limited, Bicester, Oxon.

Cover design by Iguana Creative Design

Printed in Great Britain at the University Press, Cambridge

British Library cataloguing in publication data

A catalogue record for this book is available from the British Library.

ISBN 0 521 40882 2

Contents

Introduction to 16–19 Mathematics (v)

Why 16–19 Mathematics?
Structure of the courses
Material

Introduction to the unit (ix)

1 **Circular motion** 1

   Discussion point commentary
   Tasksheet commentary

2 **Work and kinetic energy** 5

   Discussion point commentaries
   Tasksheet commentaries

3 **Using scalar products** 13

   Discussion point commentaries
   Tasksheet commentaries

4 **Potential energy** 21

   Discussion point commentaries
   Tasksheet commentaries

5 **Modelling circular motion** 29

   Discussion point commentaries
   Tasksheet commentaries

Introduction to 16–19 Mathematics

Nobody reads introductions and nobody reads teachers' guides, so what chance does the introduction to this Unit Guide have? The least we can do is to keep it short! We hope that you will find the discussion point and tasksheet commentaries and ideas on presentation and enrichment useful.

The School Mathematics Project was founded in 1961 with the purpose of improving the teaching of mathematics in schools by the provision of new course materials. SMP authors are experienced teachers and each new venture is tested by schools in a draft version before publication. Work on *16–19 Mathematics* started in 1986 and the pilot of the course has been used by over 30 schools since 1987.

Since its inception the SMP has always offered an 'after sales service' for teachers using its materials. If you have any comments on *16–19 Mathematics*, or would like advice on its use, please write to:

> 16–19 Mathematics
> The SMP Office
> The University
> Southampton SO9 5NH

Why 16–19 Mathematics?

A major problem in mathematics education is how to enable ordinary mortals to comprehend in a few years concepts which geniuses have taken centuries to develop. In theory, our view of how to pass on this body of knowledge effectively and pleasurably has changed considerably; but no great revolution in practice has been seen in sixth-form classrooms generally. We hope that in this course, the change in approach to mathematics teaching embodied in GCSE schemes will be carried forward. The principles applied in the course are appropriate to this aim.

- Students are actively involved in developing mathematical ideas.
- Premature abstraction and over-reliance on algorithms are avoided.
- Wherever possible, problems arise from, or at least relate to, everyday life.
- Appropriate use is made of modern technology such as graphic calculators and microcomputers.
- Misunderstandings are confronted and acted upon.

By applying these principles and presenting material in an attractive way, A level mathematics is made more accessible to students and more meaningful to them as individuals. The *16–19 Mathematics* course is flexible enough to provide for the whole range of students who obtain at least a grade C at GCSE.

INTRODUCTION TO 16–19 MATHEMATICS

Structure of the courses

The A and AS level courses have a core-plus-options structure. Details of the full range of possibilities, including A and AS level *Further Mathematics* courses, may be obtained from the Joint Matriculation Board, Manchester M15 6EU.

For the A level course *Mathematics (Pure with Applications)*, students must study eight core units and a further two optional units. The structure diagram below shows how the units are related to each other. Other optional units are being developed to give students an opportunity to study aspects of mathematics which are appropriate to their personal interests and enthusiasms.

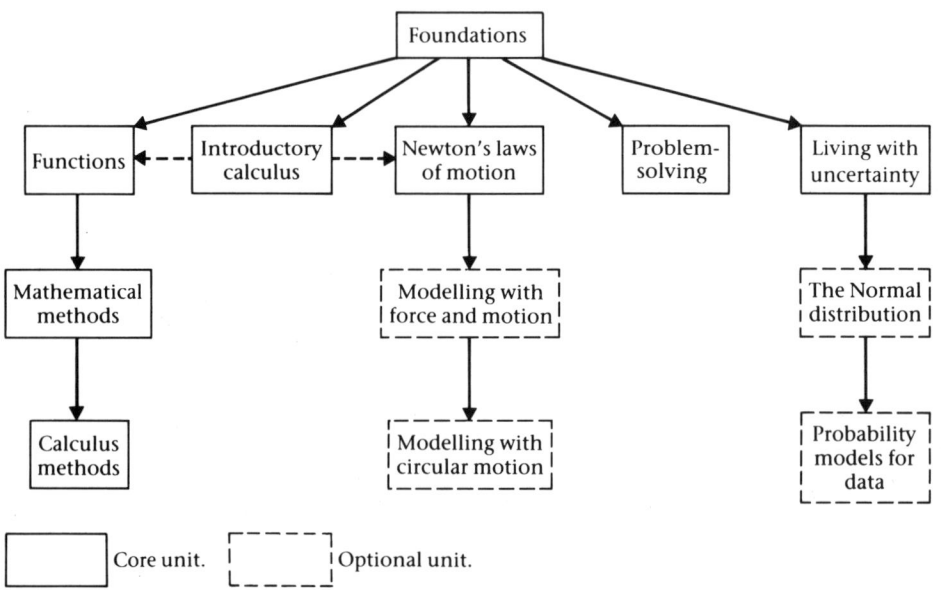

→ The *Foundations* unit should be started before or at the same time as any other core unit.

--→ Any of the other units can be started at the same time as the *Foundations* unit. The second half of *Functions* requires prior coverage of *Introductory calculus*. *Newton's laws of motion* requires calculus notation which is covered in the initial chapters of *Introductory calculus*.

For the AS level *Mathematics (Pure with Applications)* course, students must study *Foundations*, *Introductory calculus* and *Functions*. Students must then study a further two applied units.

Material

In traditional mathematics texts the theory has been written in a didactic manner for passive reading, in the hope that it will be accepted and understood – or, more realistically, that the teacher will supply the necessary motivation and deal with problems of understanding. In marked contrast, *16–19 Mathematics* adopts a questing mode, demanding the active participation of students. The textbooks contain several new devices to aid an active style of learning.

- Topics are opened up through **group discussion points**, signalled in the text by the symbol

  and enclosed in rectangular frames. These consist of pertinent questions to be discussed by students, with guidance and help from the teacher. Commentaries for discussion points are included in this unit guide.

- The text is also punctuated by **thinking points**, having the shape

  and again containing questions. These should be dealt with by students without the aid of the teacher. In facing up to the challenge offered by the thinking points it is intended that students will achieve a deeper insight and understanding. A solution within the text confirms or modifies the student's response to each thinking point.

- At appropriate points in the text, students are referred to **tasksheets** which are placed at the end of the relevant chapter. A tasksheet usually consists of a self-contained piece of work which is used to investigate a concept prior to any formal exposition. In many cases, it takes up an idea raised in a discussion point, examining it in more detail and preparing the way for formal treatment. There are also **extension tasksheets** (labelled by an E), for higher attaining students, which investigate a topic in more depth and **supplementary tasksheets** (labelled by an S) which are intended to help students with a relatively weak background in a particular topic. Commentaries for all the tasksheets are included in this unit guide.

The aim of the **exercises** is to check full understanding of principles and give the student confidence through reinforcement of his or her understanding.

Graphic calculators/microcomputers are used throughout the course. In particular, much use is made of graph plotters. The use of videos and equipment for practical work is also recommended.

As well as the textbooks and unit guides, there is a *Teacher's resource file*. This file contains:

- review sheets which may be used for homework or tests;
- datasheets;
- technology datasheets which give help with using particular calculators or pieces of software.

Introduction to the unit (for the teacher)

Examples of circular motion are all around us and catch the imagination of students. Fun-fair rides, high speed cornering in speedway races, even the humble spin dryer, all provide interesting and accessible situations that can be modelled mathematically. This unit develops the concepts introduced in *Modelling with force and motion*. Chapters 1 and 5 look at applications in the real world, while the contents of chapters 2, 3 and 4 are used to give the students the tools they will need. The concepts developed are kinetic energy, potential energy (both gravitational and elastic), the work done by a force, impulse, perfectly elastic collisions and motion in both horizontal and vertical circles.

The production of an extended piece of coursework is an integral part of the unit. Most students should be able to decide on their own project. Many of the fairground rides such as those discussed in chapter 5 are suitable. Situations used as examples or for exercise questions may also be appropriate starting points. A possible investigation may be suggested by something the student has seen. For example, a knowledge of the principles of circular motion and work and energy were obviously vital in the design of the airport chute shown in the photograph.

Some additional notes on the chapters may prove helpful.

Chapter 1

This chapter revises the work on motion in a horizontal circle and introduces the conical pendulum. The practical in the discussion point enables students to formalise their ideas on the possible differences between motion in a horizontal and vertical plane. If students' misconceptions are confronted at this stage, then their following work will have a sound basis. All the students should feel the pull of the string on their hands as the bob whirls round. It is a false economy of time to only deal with the situation theoretically because students think that they know what will happen.

Chapter 2

This shows students how to analyse the motion of an object under a variable force and derives the work – energy equation. For questions which can only be solved by numerical methods, students are shown how to use both (time, force) graphs and (distance, force) graphs to estimate the change in momentum, work done and the force acting. The last section deals with collisions for which kinetic energy is conserved and introduces Newton's experimental law for a perfectly elastic collision.

Chapter 3

Much of this chapter deals with the rules of scalar products. Students who have already met the theory in another unit will find this easy, but there are many new ideas necessary for understanding the work done by a force. The motion studied in column vector form is all in two dimensions but the general concepts are obviously equally applicable to three dimensions.

Chapter 4

This provides much of the groundwork needed for a true understanding of motion in a vertical circle under gravity. Care has been taken to ensure that the student is not left with any misconceptions about potential energy and conservation of mechanical energy. Many of the ideas here, both in the examples and the exercises, can form the basis of an extended investigation.

Chapter 5

The work done here should enable students to use the concepts of work and energy to explore motion in a vertical circle in real situations. Although most of those discussed are those from the theme park or fun-fair, *The skier* enables the student to explore the motion during a downhill race when the skier needs to maintain contact with the ground while travelling as fast as possible over a convex slope.

The time needed for the text as a whole is such that there should be plenty of time available for the students' own projects where they can use and apply all they have learnt to a real problem. Students should be encouraged to test the theories that they produce either by using a practical experiment on a model, by using video, or by experiencing the real thing as in a fun-fair ride. This is an important part of validation and should not be omitted.

Equipment needed

Masses, string, a loop-the-loop track and a ball or marble, a motor driving a conical pendulum, trucks and track for impact experiments, springs and rubber cord, a large cylinder or circular cake tin, a turntable with a banking device.

The track, the conical pendulum and the turntable and motor are available as part of the **Leeds Mechanics Kit**, available from Unilab. The rest of the equipment needed can be found in the **16–19 Mechanics Kit**, also from Unilab.

The students may need additional equipment for validation during their own investigations but this is unlikely to involve anything that cannot be found either at home or in one of the two kits mentioned above.

Tasksheets and resources

1 Circular motion

1.1 Modelling horizontal circular motion
    Tasksheet 1 – The chair-o-plane
1.2 Investigating vertical circular motion

2 Work and kinetic energy

2.1 Areas under graphs
    Tasksheet 1 – The moving car
2.2 Speed and distance
2.3 Work done by a variable force
2.4 Collisions
    Tasksheet 2E – Sporting collisions

3 Using scalar products

3.1 Work done in two dimensions
3.2 The scalar product
    Tasksheet 1 – Investigating the scalar product
3.3 Using column vectors
3.4 Work done by several forces
    Tasksheet 2 – Work done by several forces
3.5 Variable forces

4 Potential energy

4.1 Gravitational potential energy
4.2 Conserving energy
    Tasksheet 1 – A flying ball
4.3 Elastic potential energy
    Tasksheet 2 – Tension
4.4 Conserving mechanical energy

5 Modelling circular motion

5.1 Changing speed, changing energy
5.2 Acceleration
    Tasksheet 1 – Calculating acceleration
5.3 Losing contact
    Tasksheet 2 – The skier
5.4 Investigations

1 Circular motion

1.1 Modelling horizontal circular motion

Tie a small mass or bob to the end of a piece of string. Set the bob moving in a horizontal circle.

(a) What forces act on the bob and on the string?

(b) Are these forces constant?

(c) What do you have to do to get small circles? How do you get large circles?

(d) What force does your hand feel? Why?

(a) Two forces act on the bob, weight, **W**, and tension, **T**.

The string experiences two forces, the pull of the bob on one end and the pull of the hand on the other. The string is said to be in tension.

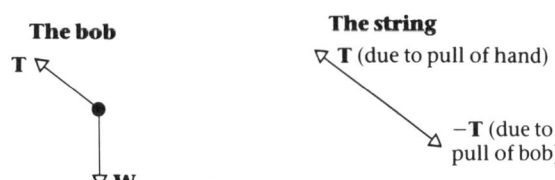

(b) Weight is a constant force; tension is not. Although the magnitude of tension may be assumed to be constant, the **direction** is not.

(c) If the angular speed of the bob is increased then the radius of the circular motion also increases. In reality, as the angular speed increases, you can feel the tension increasing. Similarly, as you decrease the angular speed and the tension decreases, then the radius of the circle decreases. You could also keep the angular speed the same but change the length of the string.

(d) Your hand feels an 'outward' and 'downward' force due to the tension in the string pulling your hand. This is equal and opposite to the 'inward' and 'upward' force you apply to the string to pull the bob around in a circle. This is an example of Newton's third law of motion.

1.2 Investigating vertical circular motion

(a) Discuss the different possibilities for the motion of the marble released at the start of the loop-the-loop shown above.

(b) Set up a track to validate your answers to (a). What is the main difference between vertical and horizontal circular motion?

(a) There seem to be three different possibilities for the motion of the marble.

  (i) The marble completes a loop safely.

  (ii) The marble falls off before it reaches the top of the loop and becomes a projectile until it lands on the track on the other side (if you are lucky).

  (iii) The marble goes part way up the loop before it rolls back down again.

(b) This can be validated by setting up the apparatus and releasing the marble from different heights. Even when the marble completes a full vertical circle, its speed varies. It goes more slowly at the top of the loop than it does at the bottom. The horizontal circular motion which you have analysed so far has always assumed constant angular speed. You will not necessarily be able to assume constant speed (or angular speed) in situations involving vertical circular motion.

However, it would be wrong to assume that constant speed is **never** found in vertical circular motion. Several fairground rides **do** have vertical circular motion **and** constant speed. It would be equally wrong to assume that constant speed is always a feature of horizontal circular motion. When you negotiate a bend in a car you do not necessarily do so at a constant speed!

You may have used expressions such as mass, weight, friction, momentum, acceleration, velocity, angular velocity, speed, displacement, distance, time taken, rate of change of kinetic energy/potential energy, work done by a force, conservation of momentum, conservation of energy, . . ., to explain or describe what was happening in the discussion point. Many of these concepts have been looked at in detail in *Newton's laws of motion* and *Modelling with force and motion*. The concepts of **work** and **energy** have not yet been dealt with on this course, although you may have come across them in other subjects. A thorough understanding of these concepts is necessary before you can develop a satisfactory model for vertical circular motion.

The chair-o-plane

TASKSHEET 1 COMMENTARY

1. As $\sin \theta \neq 0$, $T = ml\omega^2$,
   But $T \cos \theta = mg$
   $\Rightarrow ml\omega^2 \cos \theta = mg$
   $\Rightarrow \cos \theta = \dfrac{g}{l\omega^2}$ as $m \neq 0$

2. The relationship between the angle, length and angular speed, $\cos \theta = \dfrac{g}{l\omega^2}$, is independent of mass, so θ does not depend on the mass of the bob. Therefore, the heavier bob should swing out at the same angle as the lighter bob.

3. (a)

   [Graph of $\cos \theta$ against θ from 0 to 90°, starting at 1 and decreasing to 0]

   $\cos \theta = \dfrac{g}{l\omega^2}$,
   so if l is constant, as ω increases, $\cos \theta$ decreases.
   $\Rightarrow \theta$ increases
   so as ω increases, θ will also increase.

   (b) $\cos \theta = \dfrac{g}{l\omega^2}$. No matter how much you increase ω or l, the expression $\dfrac{g}{l\omega^2}$ is always positive; it can never reach zero. It follows that θ can only get very close to 90° and can never reach (or go beyond) 90°.

   (c) For constant angular speed, $l \cos \theta = \dfrac{g}{\omega^2}$, 'a constant'.
   $\Rightarrow$ As l increases, $\cos \theta$ decreases and therefore θ increases.

   (d) $l \cos \theta = \dfrac{g}{\omega^2}$ and $h = l \cos \theta \Rightarrow h = \dfrac{g}{\omega^2}$
   This relationship is independent of mass and length so you would expect h to be the same for both bobs.

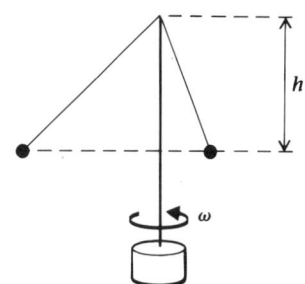

3

TASKSHEET COMMENTARY 1

4 If $\omega < \sqrt{\left(\dfrac{g}{l}\right)}$, the string remains vertical.

The bob will not start to swing out until the angular speed exceeds $\sqrt{\left(\dfrac{g}{l}\right)}$.

If you refer back to the analysis section of the tasksheet you will notice that the equation $T \sin \theta = ml\omega^2 \sin \theta$ has two possible solutions: $T = ml\omega^2$ and $\sin \theta = 0$. If $\sin \theta = 0$, then the string is vertical.

Notice that the angular speed at which a bob starts to swing out is inversely proportional to $\sqrt{l}$. When you tie two bobs to the same spindle with different lengths of string, you will find that the bob on a long string will swing out before the bob on the short string, as the angular speed increases.

5 The main difference is that the 'bob' on a chair-o-plane is suspended some distance from the axis of rotation. An analysis of the chair-o-plane will have to take this into account.

2 Work and kinetic energy

2.1 Areas under graphs

> If the mass of the car was 1200 kg, estimate the original speed of the car if it came to rest after 0.1 second.

A variable force can be approximated by a series of forces, constant over short time-intervals.

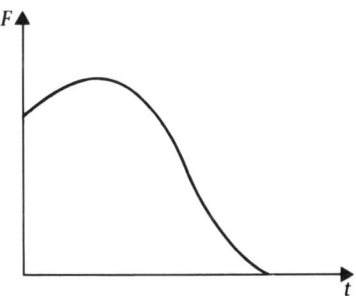

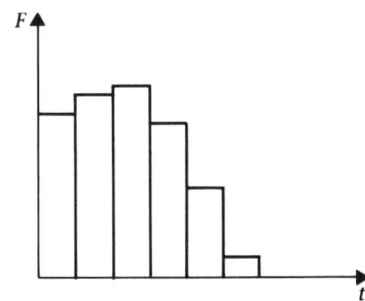

The area under the step graph measures a succession of 'changes in momentum'. To find the change in momentum of the car in the safety test it is therefore necessary to find the area under the (time, force) graph. This can be estimated by counting squares or using trapezia.

The area under the graph is approximately 17 000 N s.

So $17\,000 \approx 1200\,v$, where v is the original speed.

$$v \approx 14\,\mathrm{m\,s^{-1}}$$

2.2 Speed and distance

A series of skid tests is carried out in which a car skids to rest with its wheels locked by the brakes. The table below shows the lengths of skid marks, x metres, for various speeds, $u\,\text{km}\,\text{h}^{-1}$.

u	0	40	60	80	100
x	0	9	20	36	56

What would you expect the length of the skid marks to be for an initial speed of $120\,\text{km}\,\text{h}^{-1}$? Find x in terms of u.

Plotting a (u, x) graph with these values produces a shape as shown:

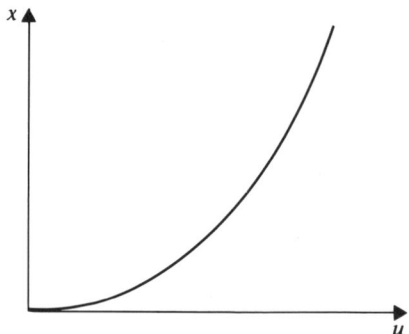

When u is doubled (from 40 to 80), x is multiplied by 4 (from 9 to 36). This suggests that x is proportional to u^2.

Trying pairs of values for u and x shows that the relationship appears to be:

$$x = \frac{u^2}{180}$$

When $u = 120$, $x = 80$.

For an initial speed of $120\,\text{km}\,\text{h}^{-1}$, the skid marks may be expected to be of length 80 metres.

2.3 Work done by a variable force

> (a) Estimate the speed of the car after it has travelled 10, 20, 30, 40 and 50 metres.
>
> (b) How would you expect the result:
>
> $Fx = \tfrac{1}{2}mv^2 - \tfrac{1}{2}mu^2$
>
> to generalise for a variable force? Justify your answer as carefully as possible.

(a) Assume, as a reasonable approximation, that the force is 3800 N during the first 10 metres, 3675 N during the next 10 metres, and so on. The work done in each 10-metre interval is then as follows:

Distance travelled (m)	0–10	10–20	20–30	30–40	40–50
Work done (J)	38 000	36 750	35 000	32 750	30 000
Total work done (J)	38 000	74 750	109 750	142 500	172 500

The additional kinetic energy equals the total work done and is 38 000 joules in the first 10 metres.

$$38\,000 = \tfrac{1}{2} \times 1000 \times v^2 - \tfrac{1}{2} \times 1000 \times 0^2$$
$$\Rightarrow v^2 = 76$$

The speed is approximately 8.7 m s^{-1}.

Similarly, the speeds after 20, 30, 40 and 50 metres are 12.2, 14.8, 16.9 and 18.6 m s^{-1}. The additional kinetic energy has been obtained by adding the areas under the step graph.

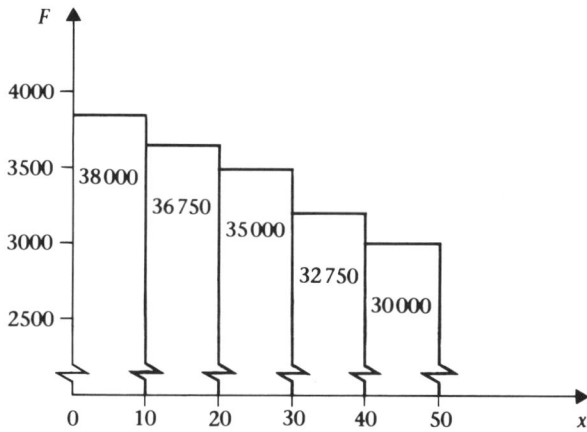

(b) You might expect a better approximation to be obtained by drawing a continuous curve through the known points and finding the area under the graph.

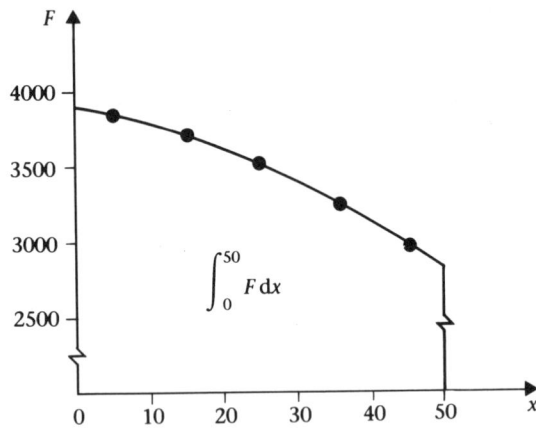

The general result is $\int F\,dx = \frac{1}{2}mv^2 - \frac{1}{2}mu^2$.

Extension – a calculus derivation of the work and energy equation.

For straight-line motion, Newton's second law can be written as:

$$F = m\frac{dv}{dt}$$
$$= m\frac{dv}{dx}\frac{dx}{dt}$$
$$= mv\frac{dv}{dx}, \quad \text{since } \frac{dx}{dt} = v$$
$$\Rightarrow \int F\,dx = \int mv\frac{dv}{dx}dx$$
$$= \int mv\,dv$$
$$= \left[\frac{1}{2}mv^2\right]_u^v$$
$$= \frac{1}{2}mv^2 - \frac{1}{2}mu^2$$

2.4 Collisions

(a) Check that momentum is conserved in all three collisions.

(b) The three collisions involve the trucks

    (i) coupling together,   (ii) using spring buffers,
    (iii) using cork buffers.

Which is which?

(c) Is kinetic energy conserved in any or all of these collisions?

(a)–(c) In all three experiments:

the momentum before collision is $mu + m \times 0 = mu$.

The kinetic energy before the collision is $\frac{1}{2}mu^2 + \frac{1}{2}m \times 0^2 = \frac{1}{2}mu^2$.

First experiment

The momentum after the collision is $m \times 0 + mu = mu$.
The momentum is conserved.

The KE after the collision is $\frac{1}{2}m \times 0^2 + \frac{1}{2}mu^2 = \frac{1}{2}mu^2$.

KE is conserved.

Second experiment

The momentum after the collision is $m \times \frac{u}{2} + m \times \frac{u}{2} = mu$.

The momentum is conserved.

The KE after the collision is $\frac{1}{2}m\left(\frac{u}{2}\right)^2 + \frac{1}{2}m\left(\frac{u}{2}\right)^2 = \frac{1}{4}mu^2$.

KE is not conserved.

Third experiment

The momentum after the collision is $m \times \frac{u}{4} + m \times \frac{3u}{4} = mu$.

The momentum is conserved.

The KE after the collision is $\frac{1}{2}m\left(\frac{u}{4}\right)^2 + \frac{1}{2}m\left(\frac{3u}{4}\right)^2 = \frac{5}{16}mu^2$.

KE is not conserved.

(b) (i) In the case where the trucks couple together, they will move with the same velocity.

It follows that the second experiment is the case of the trucks coupling.

(ii) Where spring buffers are used, one truck will be slowed to rest and the other will start to move away from it.

It follows that the first experiment is the case with the spring buffers.

(iii) The third experiment must, therefore, be the case with the cork buffers.

The moving car

TASKSHEET COMMENTARY 1

1

t (s)	1	3	7
v (m s^{-1})	$\dfrac{15}{4}$	$\dfrac{80}{9}$	$\dfrac{428}{25}$
(a) mv (Ns)	3000	7111	13696
(b) a (m s^{-1})	$\dfrac{-10t + 20}{4}$	$\dfrac{-14t + 70}{9}$	$\dfrac{-16t + 160}{25}$
	$\dfrac{10}{4}$	$\dfrac{28}{9}$	$\dfrac{48}{25}$
(c) ma (N)	2000	2489	1536

2 (a) The (time, momentum) graph is the same as the (time, velocity) graph with a vertical scale of 0 to 24×10^3 Ns replacing the scale of 0 to 30 m s^{-1}.

(b)

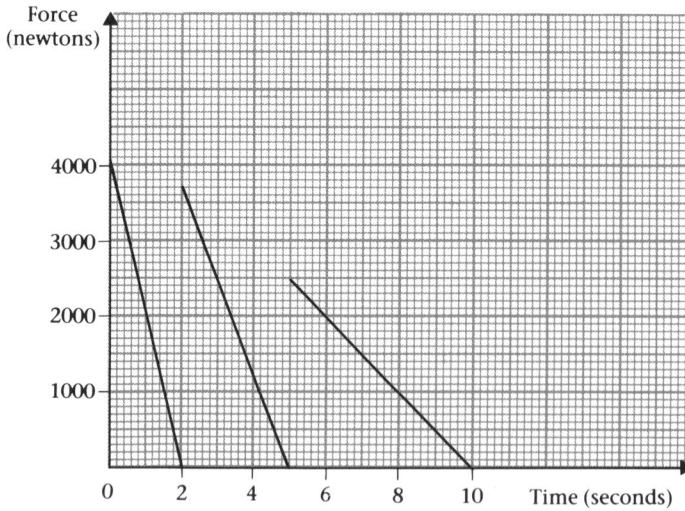

3 The area under the (time, force) graph is 16000 Ns.
It represents the increase in momentum of the car during the 10 seconds.

Sporting collisions

TASKSHEET COMMENTARY 2E

1 A good approximation to the speed of the head of the club can be obtained using the distance between successive positions of the head of the club just before the ball is struck. A similar method can be used for the ball.

4 cm represents the length of the metric rule.

1.7 cm represents the distance between the positions of the head of the club. This distance is therefore $\frac{1.7}{4}$ m.

The speed of the head of the club is $\frac{1.7}{4} \times 100 = 42.5 \, \mathrm{m\,s^{-1}}$.

2.5 cm represents the distance between successive positions of the ball in flight.

The speed of the golf ball after being struck is $\frac{2.5}{4} \times 100 = 62.5 \, \mathrm{m\,s^{-1}}$.

2 The golf ball is a very 'bouncy' object and it is not unreasonable to consider the collision between the club and the golf ball as being perfectly elastic.

The most significant difference between the mathematical model and reality is that the model relates to motion in a horizontal straight line whereas the ball is driven into the air.

3 (a) Assuming the collision is perfectly elastic, the speed of the approach = u = the speed of separation. As m separates from M, v must be greater than the speed of M and, as all velocities are in the same direction, the difference of the speeds after separation must be u. Therefore the speed of M is $v - u$.

(b) As all the velocities are taken in the same direction, this may be taken as the positive direction for momentum. The total momentum before collision is Mu and after the collision it is $M(v - u) + mv$.

Since momentum is conserved, these two expressions are equal.

(c) Since $m > 0$, $M + m > M$, and therefore:

$$\frac{2M}{M + m} < \frac{2M}{M} = 2$$

(If the divisor of a number is increased, the answer is decreased.)

TASKSHEET COMMENTARY 2E

4 The following neat method avoids the use of calculus.

$$\frac{4x}{(1+x)^2} = \frac{(1+x)^2 - (1-x)^2}{(1+x)^2} = 1 - \frac{(1-x)^2}{(1+x)^2} = 1 - \left(\frac{1-x}{1+x}\right)^2$$

Now, being a perfect square, $\left(\frac{1-x}{1+x}\right)^2 \geq 0$. The least value, 0, only occurs when $x = 1$. Therefore, the greatest value of $\frac{4x}{(1+x)^2}$ is 1, occurring when $x = 1$.

Alternatively, the graph may be plotted on a graphic calculator, or you may use calculus.

If $y = \frac{4x}{(1+x)^2}$,

$$\frac{dy}{dx} = \frac{(1+x)^2 \times 4 - 4x \times 2(1+x)}{(1+x)^4}$$

$$= \frac{(1+x)(4 + 4x - 8x)}{(1+x)^4} = \frac{4 - 4x}{(1+x)^3}$$

$\frac{dy}{dx} = 0$ when $x = 1$

x	$\to$	1	$\to$
$\frac{dy}{dx}$	+	0	−

$\Rightarrow x = 1$ gives a maximum value of $\frac{4}{2^2} = 1$

5 A 'bouncy' collision in sport, where the masses are equal, occurs in snooker.

3 Using scalar products

3.1 Work done in two dimensions

> The deck of the ship is 20 metres above sea level and the lifebelt has mass 3 kg. Assume $g = 10 \, \text{N kg}^{-1}$.
>
> Suppose the lifebelt is projected horizontally with speed:
>
> (a) $10 \, \text{m s}^{-1}$ (b) $20 \, \text{m s}^{-1}$ (c) $30 \, \text{m s}^{-1}$ (d) $u \, \text{m s}^{-1}$
>
> For each speed of projection calculate:
>
> (i) the displacement vector, **r**, of the swimmer from the deck;
>
> (ii) the change in the kinetic energy of the lifebelt from the point of projection to when it lands in the water.
>
> What does this suggest about work done on the belt during this time?
>
> Calculate the work done by gravity if the lifebelt is **dropped** from the deck into the sea.

(a) If $\mathbf{u} = \begin{bmatrix} 10 \\ 0 \end{bmatrix}$, $\mathbf{r} = \begin{bmatrix} 10t \\ -5t^2 \end{bmatrix}$ and $u = 10$

The belt hits the water when $-5t^2 = 20 \Rightarrow t = 2$

Hence $\mathbf{r} = \begin{bmatrix} 20 \\ 20 \end{bmatrix}$ and $\mathbf{v} = \begin{bmatrix} 10 \\ -20 \end{bmatrix} \Rightarrow v = \sqrt{500}$

Initial KE = 150 J and final KE = 750 J
The change in KE is 600 joules.

(b) If $\mathbf{u} = \begin{bmatrix} 20 \\ 0 \end{bmatrix}$, $\mathbf{r} = \begin{bmatrix} 40 \\ -20 \end{bmatrix}$ and change in KE = 1200 − 600
= 600 joules

(c) If $\mathbf{u} = \begin{bmatrix} 30 \\ 0 \end{bmatrix}$, $\mathbf{r} = \begin{bmatrix} 60 \\ -20 \end{bmatrix}$ and change in KE = 1950 − 1350
= 600 joules

(d) If $\mathbf{u} = \begin{bmatrix} u \\ 0 \end{bmatrix}$, $\mathbf{r} = \begin{bmatrix} 2u \\ -20 \end{bmatrix}$ and change in KE $= \dfrac{3u^2}{2} + 600 - \dfrac{3u^2}{2}$
= 600 joules

The work done is independent of the horizontal distance travelled.
The work done by gravity is $3g \times 20 = 600 \, \text{J}$.

3.2 The scalar product

A labourer has to move bricks from the ground floor to the first floor of a building. Using a hod to carry the bricks he walks up a 4 metre slope of angle 15° and then climbs a 5 metre ladder inclined at 70° attached to the scaffolding. The force he exerts on the bricks is modelled as a constant vertical force of magnitude 500 newtons.

What is the work done by the 500 newton force acting on the bricks?

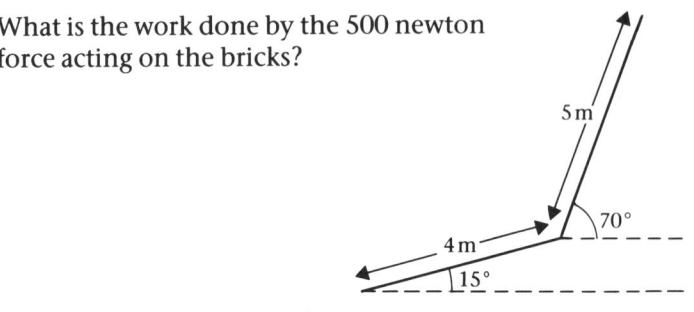

Total work done = 500 × 4 cos 75° + 500 × 5 cos 20°
= 518 + 2349 = 2867 joules

The work done can also be found by the following method. The bricks are raised to a total height of:

4 sin 15° + 5 sin 70° = 5.734 metres

The work done is therefore 500 × 5.734 = 2867 joules

3.3 Using column vectors

(a) Use this result to check the answer to the question posed in the thinking point above.

(b) What are the advantages/disadvantages of this method for calculating the work done by a force?

(a) Work done = 3 × 0 + 3 × 4 = 12 joules

(b) The magnitudes and angles do not need to be calculated to find the work done. This means that the actual arithmetic is simpler and there is less likelihood of error.

3.4 Work done by several forces

In the example above:

(a) calculate the work done by each of the three forces.

(b) What is the resultant force acting on the load?

(c) What work is done by the resultant force?

(d) What is the total work done on the load?

(e) Explain what you have found.

(a) Work done by force $\mathbf{R} = \begin{bmatrix} -40 \\ 160 \end{bmatrix} \cdot \begin{bmatrix} 28 \\ 7 \end{bmatrix}$

$= -1120 + 1120$
$= 0$ joules

Work done by force $\mathbf{T} = \begin{bmatrix} 40 \\ 10 \end{bmatrix} \cdot \begin{bmatrix} 28 \\ 7 \end{bmatrix}$

$= 1120 + 70$
$= 1190$ joules

Work done by force $\mathbf{W} = \begin{bmatrix} 0 \\ -170 \end{bmatrix} \cdot \begin{bmatrix} 28 \\ 7 \end{bmatrix}$

$= -1190$ joules

(b) Resultant force $= \begin{bmatrix} -40 \\ 160 \end{bmatrix} + \begin{bmatrix} 40 \\ 10 \end{bmatrix} + \begin{bmatrix} 0 \\ -170 \end{bmatrix}$

$= \begin{bmatrix} 0 \\ 0 \end{bmatrix}$ newtons

(c) Work done by resultant force $= \begin{bmatrix} 0 \\ 0 \end{bmatrix} \cdot \begin{bmatrix} 28 \\ 7 \end{bmatrix}$

$= 0$ joules

(d) Total work done $= 0 + 1190 + -1190$
$= 0$ joules

(e) The work done by the resultant force is equal to the sum of the work done by the individual forces.

3 USING SCALAR PRODUCTS

3.5 Variable forces

Not all forces acting on an object do work. What can you say about forces which do no work?

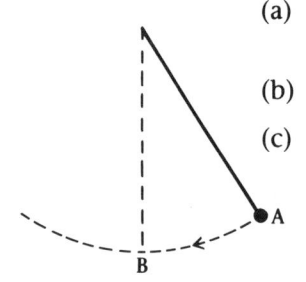

(a) What are the forces acting on the bob (mass m) of a pendulum?

(b) Are these forces constant?

(c) What work does each force do when the pendulum moves from A to B?

If a force does no work then its direction must be at right angles to the direction of motion.

(a) The forces acting on the bob of the pendulum are:

(i) the tension in the string, **T**;

(ii) the weight of the bob, **W**.

(b) The weight remains constant, but the tension varies in both magnitude and direction.

(c) The tension force does no work as it is always perpendicular to the direction of motion.
The work done by the weight $= \mathbf{W} \cdot \mathbf{r}$

$$= \begin{bmatrix} 0 \\ -mg \end{bmatrix} \cdot \overrightarrow{AB}$$

$$\overrightarrow{AB} = \begin{bmatrix} -AN \\ -NB \end{bmatrix}$$

$$\begin{bmatrix} 0 \\ -mg \end{bmatrix} \cdot \begin{bmatrix} -AN \\ -NB \end{bmatrix} = mg\,(NB)$$

So the work done by the weight $= W \times$ vertical distance AB

Investigating the scalar product

TASKSHEET COMMENTARY 1

1. (a) Work done = 200 × 4
   = 800 J

   (b) Work done = 200 × 4 cos 30°
   = 400 √3 J

   (c) Work done = 200 × 4 cos 60°
   = 400 J

   (d) Work done = 200 × 4 cos 90°
   = 0 J

2. Both forms give the correct answer for the scalar product because cos 60° = cos 300°. The angle must be measured between the two vectors drawn from a common point. The vectors must therefore be thought of as:

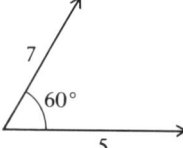   and not as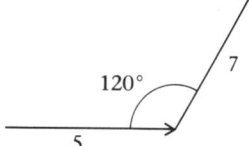

3. (a) **R** is perpendicular to the displacement so the work done is $R \times 5 \times \cos 90° = 0$.

   (b) Work done by the weight = 35 × 5 × cos 120°
   = −87.5 J

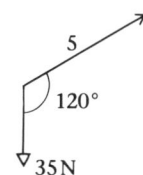

4. (a) Work done by the force = 30 × 2 = 60 J

   (b) Work done by the force = 30 × 2 = 60 J

   (c) In each case the force is in the same direction as the displacement and so the work done is positive.

   (d) $(-\mathbf{p}) \cdot (-\mathbf{q}) = -p \times -q \times \cos \theta$
   $= pq \cos \theta$
   $= \mathbf{p} \cdot \mathbf{q}$

5. (a) Work done = 7 × 7
   = 49 J

   (b) $\mathbf{p} \cdot \mathbf{p} = p \times p \times \cos 0° = p^2$

3 USING SCALAR PRODUCTS

TASKSHEET
COMMENTARY 1

6 (a) Work done by the force = 500 × 5 sin 25° = 1057 J

(b) The new distance is 15 m.
Work done by the force = 500 × 15 sin 25° = 3171 J
Three times as much work is done.

(c) The new force = 1000 N
The work done = 1000 × 5 sin 25°
= 2 × 1057 J   i.e. the work done is doubled.

(d) The new force = 1000 N
The new distance = 15 m
The work done = 1000 × 15 sin 25°
= 6 × 1057 J
The work done is six times as great.

(e) $(k\mathbf{p}) \cdot (l\mathbf{q}) = kp \times lq \times \cos\theta$
$= klpq \cos\theta$
$= kl\,(\mathbf{p} \cdot \mathbf{q})$

7 Total work done = 100 × 4 sin 25° + 100 × 6 sin 10°
= 273 J

8 (a)

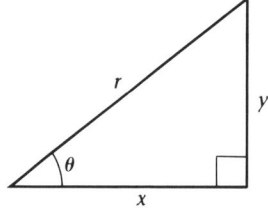

From the diagram in question 7 it follows that:

$x = 6\cos 10° + 4\cos 25° = 9.53\,\text{m}$

$y = 6\sin 10° + 4\sin 25° = 2.73\,\text{m}$

$\Rightarrow r = \sqrt{(x^2 + y^2)} = 9.91\,\text{m}$

and $\theta = \tan^{-1}\dfrac{y}{x} = 16.0°$

(b) $Wr\cos(90° - \theta) = 100 \times 9.91 \times \cos 74°$
= 273 joules

(c)

The work done by gravity in question 7 is the same as in (b) above. This shows that
$\mathbf{W} \cdot \mathbf{r}_1 + \mathbf{W} \cdot \mathbf{r}_2 = \mathbf{W} \cdot \mathbf{r}$
$= \mathbf{W} \cdot (\mathbf{r}_1 + \mathbf{r}_2)$

$\mathbf{r}_1 + \mathbf{r}_2 = \mathbf{r}$

9 Work done by force up the ramp 5 m long = 500 × 5 sin 30°
= 1250 J

Work done by force up the ramp 4 m long = 500 × 4 sin 48.6°
= 1500 J

The difference in the work done by the force is 1500 − 1250 = 250 J

18

Work done by several forces

TASKSHEET COMMENTARY 2

1 (a) Work done by the force of $\begin{bmatrix} 2 \\ 3 \end{bmatrix}$ newtons is $\begin{bmatrix} 2 \\ 3 \end{bmatrix} \cdot \begin{bmatrix} 5 \\ 7 \end{bmatrix} = 31$ J

Work done by the force of $\begin{bmatrix} 4 \\ -1 \end{bmatrix}$ newtons is $\begin{bmatrix} 4 \\ -1 \end{bmatrix} \cdot \begin{bmatrix} 5 \\ 7 \end{bmatrix} = 13$ J

Work done by the force of $\begin{bmatrix} -3 \\ -2 \end{bmatrix}$ newtons is $\begin{bmatrix} -3 \\ -2 \end{bmatrix} \cdot \begin{bmatrix} 5 \\ 7 \end{bmatrix} = -29$ J

(b) The resultant of the three forces is $\begin{bmatrix} 3 \\ 0 \end{bmatrix}$ newtons.

(c) Work done by the resultant is $\begin{bmatrix} 3 \\ 0 \end{bmatrix} \cdot \begin{bmatrix} 5 \\ 7 \end{bmatrix} = 15$ J

(d) The total work done by all the forces is equal to the work done by the resultant force.

2 The work done by the resultant should be equal to the sum of the work done by each of the component forces.

3 (a) $\begin{bmatrix} 3 \\ 2 \end{bmatrix} \cdot \begin{bmatrix} 6 \\ -3 \end{bmatrix} = 18 - 6 = 12$ J

$\begin{bmatrix} 4 \\ -8 \end{bmatrix} \cdot \begin{bmatrix} 6 \\ -3 \end{bmatrix} = 24 + 24 = 48$ J

$\begin{bmatrix} 2 \\ -6 \end{bmatrix} \cdot \begin{bmatrix} 6 \\ -3 \end{bmatrix} = 12 + 18 = 30$ J

$\begin{bmatrix} 3 \\ 6 \end{bmatrix} \cdot \begin{bmatrix} 6 \\ -3 \end{bmatrix} = 18 - 18 = 0$ J

(b) The resultant force $= \begin{bmatrix} 3 \\ 2 \end{bmatrix} + \begin{bmatrix} 4 \\ -8 \end{bmatrix} + \begin{bmatrix} 2 \\ -6 \end{bmatrix} + \begin{bmatrix} 3 \\ 6 \end{bmatrix} = \begin{bmatrix} 12 \\ -6 \end{bmatrix}$ newtons

Work done $\begin{bmatrix} 12 \\ -6 \end{bmatrix} \cdot \begin{bmatrix} 6 \\ -3 \end{bmatrix} = 72 + 18 = 90$ J

This equals the total work done by all the forces acting.

i.e. $12 + 48 + 30 + 0 = 90$

3 USING SCALAR PRODUCTS

TASKSHEET COMMENTARY 2

(c) It is perpendicular to the displacement.

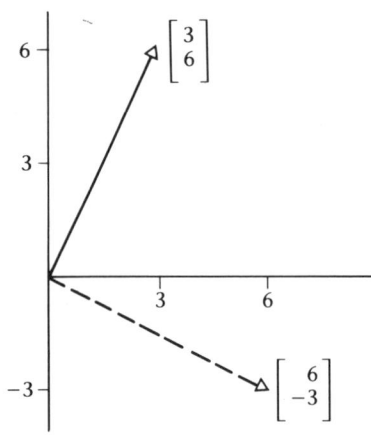

4 The resultant force is $\begin{bmatrix} 22 \\ -4 \end{bmatrix} + \begin{bmatrix} -6 \\ -8 \end{bmatrix} = \begin{bmatrix} 16 \\ -12 \end{bmatrix}$.

The magnitude of the vector $\begin{bmatrix} 4 \\ -3 \end{bmatrix}$ is 5 units.

Therefore the displacement $= 6 \begin{bmatrix} 4 \\ -3 \end{bmatrix}$ cm

$= \dfrac{6}{100} \begin{bmatrix} 4 \\ -3 \end{bmatrix}$ metres

The work done $= \dfrac{6}{100} \begin{bmatrix} 4 \\ -3 \end{bmatrix} \cdot \begin{bmatrix} 16 \\ -12 \end{bmatrix} = 6$ joules

20

4 Potential energy

4.1 Gravitational potential energy

(a) What is the total work done?

(b) What forces act on the pencil?

(c) What is the work done by each force?

(a) The pencil is stationary at both positions, so there is no change in KE.
Thus the total work done is zero.

(b) The forces acting on the pencil are the contact force between you and the pencil and the gravitational force (weight) acting downwards.

(c) Work done by the weight $= -mgh$

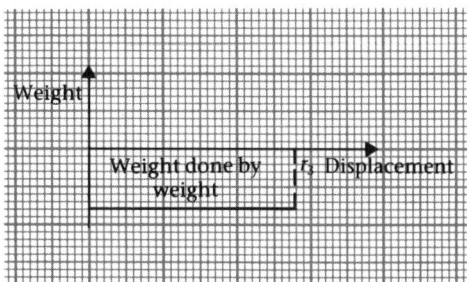

The total work is done is zero.
So work done by the hand $= mgh$

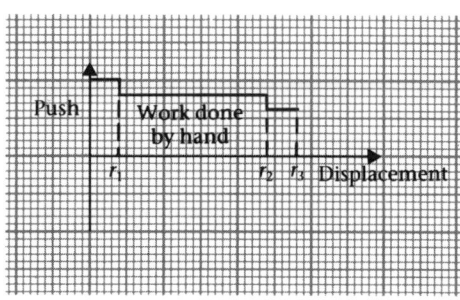

21

4 POTENTIAL ENERGY

4.2 Conserving energy

(a) How much work is done by each force acting on the child as she goes from A to B?

(b) With what speed does the child enter the water?

(c) What is the PE of the child at A? What is the KE of the child at A?

(d) What are the PE and KE of the child at B?

(e) Is energy conserved?

(f) Is energy conserved if a constant frictional force of 100 newtons opposes motion?

(a) There are two forces on the child, **N** and **W**.
N is the contact force from the chute and it acts at 90° to the direction of motion. Since the distance moved in the direction of **N** is zero, the work done by **N** on the child is zero.
W is the weight of the child, i.e. it is the gravitational attraction of the Earth on the child.

$$W = 40\,\text{kg} \times 10\,\text{N}\,\text{kg}^{-1} = 400 \text{ newtons}$$

W acts vertically and the distance moved in this direction is 3 metres.
The work done by **W** on the child = 400 newtons × 3 metres
= 1200 N m (or J)

(b) Work done = change of KE
$$1200 = \tfrac{1}{2}mv^2 - \tfrac{1}{2}mu^2$$
$$= 20v^2 - 20 \times 3^2$$
⇒ $60 = v^2 - 9$
⇒ $v^2 = 69$
⇒ $v = 8.3\,\text{m}\,\text{s}^{-1}$ = final speed of child

(c) PE at A (the top) = mgh
= 40 × 10 × 3
= 1200 J

KE at A = $\tfrac{1}{2}mu^2$
= $\tfrac{1}{2}$ × 40 × 3²
= 180 J

(d) PE at B = mgh
$$= 40 \times 10 \times 0$$
$$= 0 \text{ J}$$

KE at B = $\frac{1}{2}mv^2$
$$= \frac{1}{2} \times 40 \times 8.3^2$$
$$= 20 \times 69$$
$$= 1380 \text{ J}$$

(e) Initial (PE + KE) at A = 1200 + 180
$$= 1380 \text{ J}$$

Final (PE + KE) at B = 0 + 1380
$$= 1380 \text{ J}$$

Energy is conserved because the only force doing work is gravity. You assumed conservation of energy when calculating the speed of the child in part (b).

(f) If a constant frictional force of 100 newtons acts in a direction parallel to the chute then the work done against friction is:

$$100 \sqrt{(3^2 + 5^2)} = 100 \sqrt{34} = 583 \text{ J}$$

The amount of energy is dissipated from the system as heat and so the child starts with 1380 J at the top but has only 1380 − 583 = 797 J of energy at the bottom. Since this is in the form of KE, the child's speed at B is now:

$$\sqrt{\left(\frac{2 \times 797}{40}\right)} = 6.3 \text{ m s}^{-1}$$

4 POTENTIAL ENERGY

4.3 Elastic potential energy

If a spring with spring constant k is stretched (or compressed) a distance x beyond its natural length, show that the elastic potential energy of the spring is:

$$\frac{kx^2}{2}$$

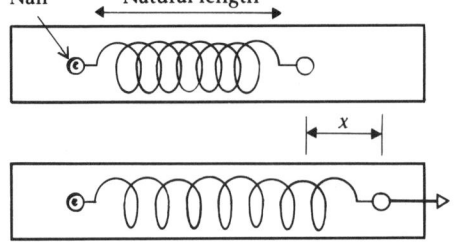

A spring is anchored by a nail to a strip of wood.

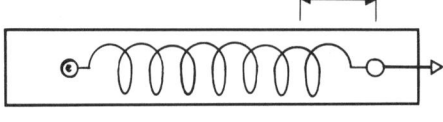

A force extends the spring a distance x.

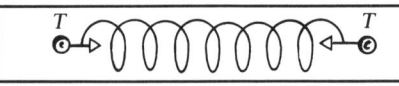

This end is also anchored with a nail, and the force is removed. The tension in the spring is $T = kx$.

In this situation there is a store of potential energy. The tension in the spring has the potential to do some work if, for example, one end is connected to an object which is free to move when the nail is removed.

If this were to happen, then the tension would decrease linearly from kx to zero as the displacement of the object increases from zero to x.

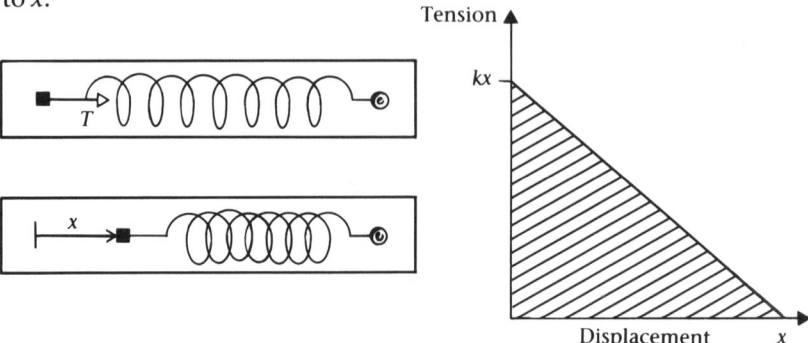

Work done by tension = area under the graph
$$= \frac{kx^2}{2}$$

The tension in the spring has the potential to do $\dfrac{kx^2}{2}$ joules of work.

4.4 Conserving mechanical energy

Set a mass on the end of a spring oscillating.

(a) Time one complete oscillation.

(b) Draw a rough sketch of the (time, KE) graph for one complete oscillation.

(c) Draw rough sketches of the graphs of the gravitational PE of the mass and of the elastic PE of the spring against the time for one complete oscillation.

(d) What can you say about the total mechanical energy of the system?

(a) Here are some techniques to help you achieve accuracy when timing oscillations.

(i) Time ten oscillations (or more if possible) to increase the length of time you are measuring. Then divide to give the period of one oscillation.

(ii) When counting the oscillations, say 'nought' as you start the watch. Then stop the watch as you say 'ten'. (If you say 'one' when starting, you will have timed only nine oscillations!)

(iii) Set up a timing mark (an arrow on a vertical card) to show the rest position of the mass.
Each complete oscillation is measured from when the mass passes the mark, going up, to the next time it passes the mark on the upward journey.

(b)
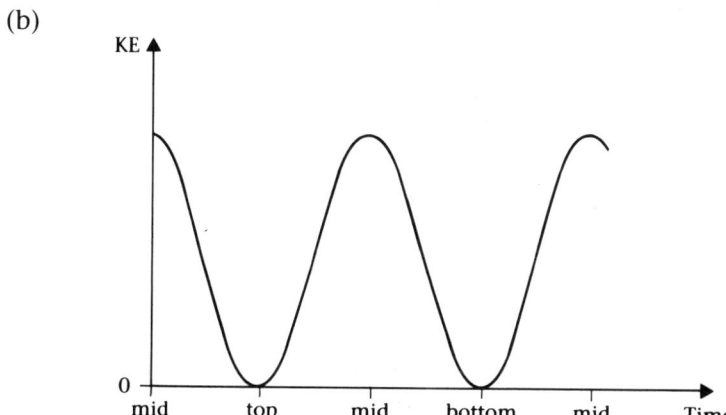

(c)

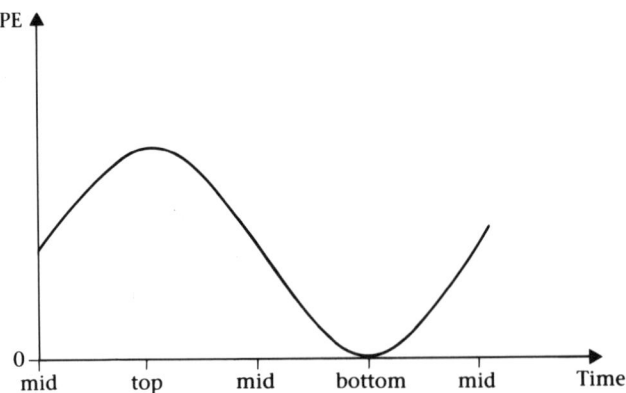

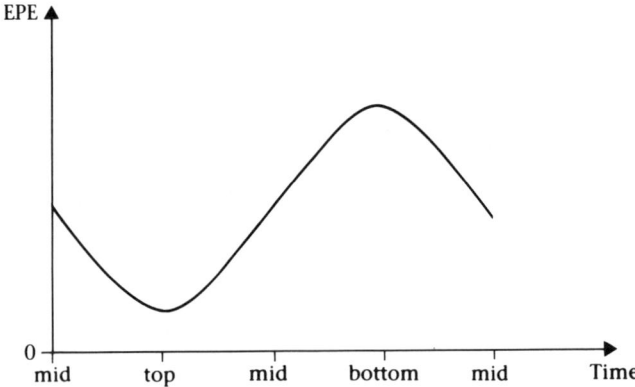

Note that the EPE does not go to zero because the spring does not become completely slack at the top.

(d) The total mechanical energy of the system appears to have been conserved. However, over a long time interval the spring will stop oscillating, showing that some energy is lost during each oscillation.

A flying ball

TASKSHEET COMMENTARY 1

Work done = change in kinetic energy
If you assume that the weight is the only force to do work, then the work done each metre as the ball rises is $-mg \times 1$.
Work done = -0.5
Initial KE = $\frac{1}{2}mv^2 = 0.025 \times 10^2 = 2.5$

Height above point of projection (metres)	KE (joules)	PE relative to point of projection (joules)
0	2.5	0
1	2.0	0.5
2	1.5	1.0
3	1.0	1.5
4	0.5	2.0
5	0	2.5

1 The kinetic energy of the ball is zero when it has risen 5 metres. Its velocity is zero, so it will not rise any higher.

2 The sum of the potential and kinetic energies is 2.5 J. Energy appears to be conserved.

3 As the kinetic energy decreases, the potential energy increases.

4 If a tennis ball is used, some work will be done against air resistance. The total work done each metre that the ball rises is 0.625 J.

Height above point of projection (metres)	KE (joules)	PE (joules)	KE + PE (joules)
0	2.500	0	2.500
1	1.875	0.5	2.375
2	1.250	1.0	2.250
3	0.625	1.5	2.125
4	0	2.0	2.000

5 Mechanical energy is converted into heat and sound energy.

Tension

TASKSHEET COMMENTARY 2

1. Various springs will give different graphs.
   As you saw in *Modelling with force and motion*, the likely graphs are as shown:

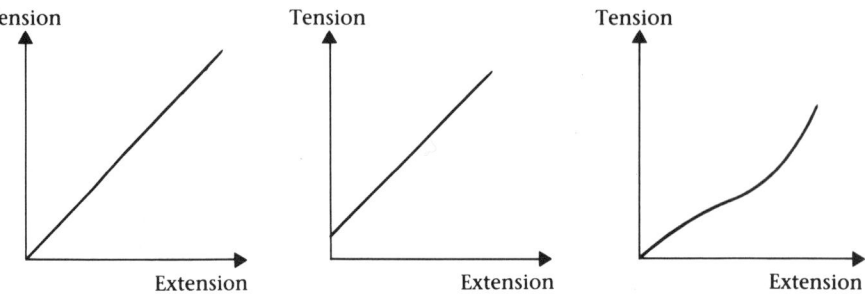

   However, most springs will obey Hooke's law for at least part of their extension.

2. The gradient of the line of best fit, passing through the origin, will give an estimation of *k*, the spring constant.

5 Modelling circular motion

5.1 Changing speed, changing energy

(a) In which of the situations above is it reasonable to assume conservation of energy?

(b) Describe what happens to the kinetic and potential energies of the marble as it moves along the track.

(c) How would using a toy car rather than a marble affect what happens?

Set up the apparatus to validate your conjectures.

(a) For the conker, some energy will be lost due to the effects of air resistance and friction at the support where the string is held. If both of these effects are small, then it is reasonable to assume that energy is conserved.

For the marble, you would expect the effect of air resistance to be negligible. Furthermore, no energy is lost because of friction, provided the marble 'rolls without slipping.' Negligible energy is therefore lost when the marble rolls around the cylinder. When the marble goes around the loop, some slipping does occur on the steep parts of the loop and therefore some energy is lost. Additional energy is lost when the marble goes over joins in the track. However, the total energy loss is likely to be sufficiently small for conservation of energy to be a reasonable assumption.

(b) When the marble is released from rest it has potential energy but no kinetic energy. As it rolls down the track its kinetic energy increases at the expense of its potential energy and kinetic energy reaches its maximum value at the bottom of the loop. When climbing the loop its potential energy increases and kinetic energy decreases.

(c) When a car is used there are additional friction losses due to the presence of internal friction acting on the axles. Experiments show that the energy lost is far greater than for the marble and it is **not** reasonable to assume that energy is conserved.

5 MODELLING CIRCULAR MOTION

5.2 Acceleration

(a) What forces are acting on the conker?

(b) Does the resultant force act towards the centre of circular motion?

(a) The forces acting on the conker are its weight, mg, and the tension, T, in the string (assuming that air resistance is negligible).

(b) 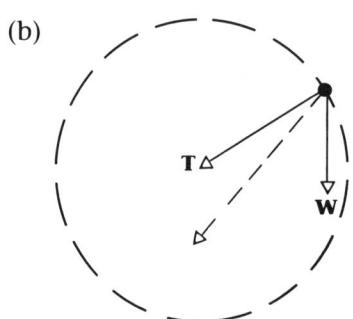 The resultant force is not directed towards the centre of circular motion unless $\theta = 0°$ or $180°$.

5.3 Losing contact

(a) What forces act on the marble as it moves along the track/cylinder?

(b) What is the condition for the marble to lose contact with the track/cylinder?

(a) If the marble is assumed to be a particle of mass m and the effects of air resistance and friction are ignored, then the forces acting are the normal contact force N and the weight mg.

(b) When the marble loses contact with the track, $N = 0$.

Calculating acceleration

TASKSHEET COMMENTARY 1

1. $\overrightarrow{OP} = \begin{bmatrix} r\cos\theta \\ r\sin\theta \end{bmatrix} = r \begin{bmatrix} \cos\theta \\ \sin\theta \end{bmatrix}$

   $\begin{bmatrix} \cos\theta \\ \sin\theta \end{bmatrix}$ is therefore in the direction $\overrightarrow{OP}$. Its length is $\sqrt{(\sin^2\theta + \cos^2\theta)} = 1$.

2. $\mathbf{r} = r \begin{bmatrix} \cos\theta \\ \sin\theta \end{bmatrix}$

   $\mathbf{v} = \dfrac{d\mathbf{r}}{dt} = r \begin{bmatrix} -\sin\theta \cdot \dot\theta \\ \cos\theta \cdot \dot\theta \end{bmatrix} = r\dot\theta \begin{bmatrix} -\sin\theta \\ \cos\theta \end{bmatrix}$

   $\begin{bmatrix} -\sin\theta \\ \cos\theta \end{bmatrix}$ is a unit vector and so $\mathbf{v}$ has magnitude $r\dot\theta$.

3. (a) Acceleration $\mathbf{a} = \dfrac{d\mathbf{v}}{dt} = r \begin{bmatrix} -\sin\theta \cdot \ddot\theta - \cos\theta \cdot \dot\theta^2 \\ \cos\theta \cdot \ddot\theta - \sin\theta \cdot \dot\theta^2 \end{bmatrix}$

   $\Rightarrow \mathbf{a} = -r\dot\theta^2 \begin{bmatrix} \cos\theta \\ \sin\theta \end{bmatrix} + r\ddot\theta \begin{bmatrix} -\sin\theta \\ \cos\theta \end{bmatrix}$

   (b) $\begin{bmatrix} -\sin\theta \\ \cos\theta \end{bmatrix}$ has length $\sqrt{(\sin^2\theta + \cos^2\theta)} = 1$ and

   $\begin{bmatrix} \cos\theta \\ \sin\theta \end{bmatrix} \cdot \begin{bmatrix} -\sin\theta \\ \cos\theta \end{bmatrix} = -\cos\theta\sin\theta + \cos\theta\sin\theta = 0$

   Therefore $\begin{bmatrix} -\sin\theta \\ \cos\theta \end{bmatrix}$ is a unit vector in a direction perpendicular to OP.

   (c) Acceleration in the radial direction is directed towards O and has magnitude:

   $r\dot\theta^2 = \dfrac{(r\dot\theta)^2}{r} = \dfrac{v^2}{r}$

The skier

TASKSHEET COMMENTARY 2

1 Assumptions should include:
- The surface is a circle of radius a.
- Air resistance is negligible.
- Forces acting on the particle are gravity, mg, and the normal contact force, N.
- The particle has speed v at P and angle AOP is θ.

2 Equating the two expressions for v^2 gives:
$$ga\cos\theta = 2ga(1 - \cos\theta)$$
$$\Rightarrow \cos\theta = \tfrac{2}{3}$$
$$\theta = \cos^{-1}(\tfrac{2}{3})$$

3 The particle will lose contact when $\cos\theta = \tfrac{2}{3}$ i.e. $\theta = \cos^{-1}(\tfrac{2}{3})$. This position is a distance below A of:
$$(a - a\cos\theta) = a(1 - \tfrac{2}{3}) = \tfrac{1}{3}a$$

4E

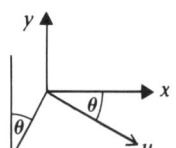

When the marble leaves the tin,
$u = \sqrt{(ga\cos\theta)} = 2.58\sqrt{a}$ tangentially.

$$\mathbf{u} = \begin{bmatrix} 2.58\sqrt{a}\cos\theta \\ 2.58\sqrt{a}\sin\theta \end{bmatrix} = \begin{bmatrix} 1.72\sqrt{a} \\ 1.92\sqrt{a} \end{bmatrix}$$

The marble travels as a projectile from P and hits the table at $\begin{bmatrix} x \\ -\tfrac{5}{3}a \end{bmatrix}$ (taking P as the origin).

$$\mathbf{r} = \begin{bmatrix} 1.72t\sqrt{a} \\ 1.92t\sqrt{a} - 5t^2 \end{bmatrix} \Rightarrow 5t^2 - 1.92t\sqrt{a} - \tfrac{5}{3}a = 0$$
$$\Rightarrow t = 0.80\sqrt{a}$$
$$\Rightarrow x \approx 1.38\,a \text{ metres}$$

In theory, it should hit the table approximately $2.13a$ metres from the point of contact of the tin and the table.